Our Endangered Planet
GROUNDWATER

Mary Hoff
and
Mary M. Rodgers

LERNER PUBLICATIONS COMPANY • MINNEAPOLIS

Dedication: To our grandmothers, Grace Darling King and Madeline MacCord Donaldson

Thanks to Hans-Olaf Pfannkuch, James E. Laib, Kerstin Coyle, Zachary Marell, and Gary Hansen for their help in preparing this book.

Words that appear in **bold** type are listed in a glossary that starts on page 59.

LIBRARY OF CONGRESS CATALOGING-IN-PUBLICATION DATA

Hoff, Mary King.
    Our endangered planet. Groundwater / Mary Hoff and Mary M. Rodgers.
      p.   cm.
    Includes bibliographical references and index.
    Summary: Describes the global uses and abuses of groundwater and suggests ways to preserve this valuable resource.
    ISBN 0-8225-2500-3 (lib. bdg.)
    1. Water, Underground—Management—Juvenile literature. 2. Water, Underground—Juvenile literature. 3. Water—pollution. 4. Water, Underground—pollution. [1. Water, Underground.]
I. Rodgers, Mary M. (Mary Madeline), 1954- . II. Title. III. Series: Our endangered planet (Minneapolis, Minn.)
TD403.H64 1991
333.91 '04—dc20
                                         90-39140
                                            CIP
                                            AC

Manufactured in the United States of America

1 2 3 4 5 6 7 8 9 10 00 99 98 97 96 95 94 93 92 91

Front cover: *Groundwater gathers on the floor of an underground limestone cave.* *Back cover:* *(Left) Using a bucket attached to a rope, a woman from Ghana, West Africa, draws water from a narrow well. (Right) Equipment helps a worker to mop up an oil spill that threatens to pollute underground water.*

All paper used in this book is of recycled material and may be recycled.

CONTENTS

OUR ENDANGERED PLANET

In the 1960s, astronauts first traveled beyond earth's protective atmosphere and were able to look back at our planet. What they saw was a beautiful globe, turning slowly in space. That image reminds us that our home planet has limits, for we know of no other place that can support life.

The various parts of our natural environment—including air, water, plants, and animals—are partners in making our planet a good place to live. If we endanger one element, the other partners are badly affected, too.

People throughout the world are working to protect and heal earth's environment. They recognize that making nature our ally and not our victim is the way to shape a common future. Because we have only one planet to share, its health and survival mean that we all can live.

One of earth's main elements is water. In fact, fresh water and salt water cover nearly three-fourths of our planet's surface. Over time, people have become more skilled in finding and using this resource to drink, to grow food, and to create energy for running machines.

An additional supply of fresh water lies under the ground. Learning about groundwater's uses and how our behavior can conserve this valuable resource will contribute to its preservation for many generations to come.

CHAPTER ONE

THE WATER BENEATH OUR FEET

When we think of water, we usually picture something we can see—a lake, rain puddles, or a glass of crystal-clear liquid. Most of the available fresh water in the world, however, is hidden where we cannot see it. This natural supply of water, called **groundwater,** lies in the earth beneath our feet.

Groundwater makes up 97 percent of all fresh water that is not trapped in the ice that forms **glaciers.** Sometimes groundwater exists in lakes and streams that flow through underground caves and tunnels.

(Left) An underground limestone cave is one place of storage for groundwater. (Right) Glaciers—slow-moving masses of ice—freeze much of our planet's fresh water.

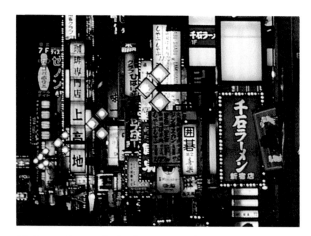

Tokyo, the capital of Japan, relies on groundwater to supply its 8.3 million people.

But the nooks and crannies within rocks and loose soil actually hold most groundwater—much like a wet sponge contains water.

Groundwater is found on every continent. This **resource**—meaning a supply of something we need—lies beneath forests, lakes, and even deserts. Although groundwater is often the most hidden source of water, it is very important to people throughout the world.

In some regions, groundwater is the only clean, fresh water available for drinking, for watering animals, and for irrigating crops. Big cities such as London, Miami, and Tokyo depend on groundwater.

Children in Botswana—a hot, dry nation in southern Africa—often must carry water from local wells to their homes.

Industries use soapy water when they loosen and clean machinery.

A hydrogeologist (a person who studies groundwater) examines a map to find out about the land surrounding a local aquifer.

This underground resource provides the water that people use in their homes. Factories need groundwater to make things that people buy. Groundwater can be used to generate electricity, to cool machinery, or to clean a finished product.

Because groundwater is so plentiful, we might believe that there is an endless supply of it. But the situation is more complicated. New technology and equipment are allowing us to locate and pump up more groundwater than in previous years.

The careless dumping of chemicals is endangering some groundwater sources.

A growing worldwide demand for water is also increasing the use of groundwater.

At the same time, strong chemicals and other harmful substances are making some groundwater unusable. In many places, people are tapping groundwater faster than it can be replaced. Others are poisoning this resource through carelessness.

Like the earth's other natural resources, groundwater is ours to use but not to use up or abuse. To protect it, each of us must first understand groundwater—where it comes from, how we use it, and what we need to do to preserve it.

HOLEY ROCKS!

Groundwater exists in the spaces, or **pores,** in the natural materials that lie beneath the ground. Most groundwater is found in **unconsolidated sediments**—layers of loose material, such as gravel and sand. But a lot of groundwater is held in rocks, too.

It is not very difficult to imagine water filling the spaces between the small particles that make up unconsolidated sediments. It is harder to believe that something as solid as rock could hold water. However, if we magnified rocks such as limestone or sandstone many times, we would see that they really are full of tiny holes. Rocks may also have cracks and larger holes that could contain groundwater. These "empty" spaces have an enormous ability to hold water that seeps down from the earth's surface.

The measure of the amount of space available to hold groundwater is called **porosity.** It is expressed as a percentage. Sand and gravel may have a porosity as high as 40 percent—meaning that 40 percent of their material is "empty" space. Some rocks, on the other hand, have a porosity of less than 1 percent.

The porosity of rocks or sediments makes a big difference in whether they are good sources of groundwater. But porosity is not the only factor. Water must also be able to travel into and between the pores—a feature called **permeability.**

Permeability is a measure of how easy it is for water to move through a certain type of soil or rock. In rocks, permeability depends on the size of the pores and on how connected they are to one another. Some rocks have many spaces (high porosity), but these spaces are separated from each other, so the overall permeability is low.

Photographed at 46 times its actual size, this piece of sandstone shows the large spaces, or pores, that exist between grains of the rock.

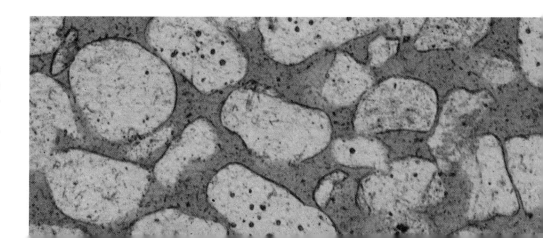

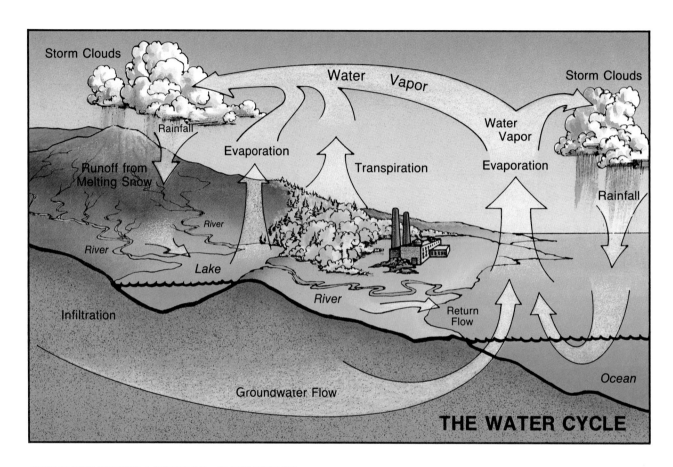

THE WATER CYCLE

GROUNDWATER'S ORIGINS

The easiest way to understand this water that we do not often see is to start with water we *do* see. Imagine a single drop of rain about to fall from a storm cloud passing over your home. We can follow that drop through the **water cycle.** It shows how water changes form and location as it travels between the air, the land, and the ocean in an endless movement.

WHO FIGURED OUT THE WATER CYCLE?

The water cycle that occurs every day on our planet was not always understood. Three Frenchmen and one Englishman are famous for explaining how the cycle worked. Bernard Palissy (1510–1589), a talented potter, also wrote articles about nature. In 1580 he stated that the water in rivers and springs came from rainfall. After Pierre Perrault (1611–1680) lost his job in Paris, France, he began to study the city's Seine River. Some scientists thought Perrault's methods were rough, but he proved that water from rain and snow caused the flow of all rivers. Using more reliable methods than Perrault did, Edme Mariotte (1620–1684) came to the same conclusions as Palissy and Perrault about how rivers received water. These three people uncovered half of the mystery of the water cycle—that rainfall supplies rivers with their water flow. But they did not explain where the rainfall comes from. The English astronomer Edmond Halley (1656–1742) solved that part of the mystery. Through experiments, Halley found that the amount of water evaporating from the Mediterranean Sea was three times as much as the water flowing into the sea from rivers. He concluded that the evaporating water—in the form of vapor—collected in storm clouds. These clouds return the water to the land and the sea as rain. Halley's work showed the constant circulation of water between oceans, the land, and the air.

Palissy

Seine River

Halley

Mariotte

The roots of plants absorb water to live. They pull the moisture through their stems and then return it to the air through their leaves.

When our drop of water hits the ground, one of several things might happen to it. If the drop lands on a surface that does not absorb water (such as a sidewalk), the drop might just sit there until the sun's heat **evaporates** it. Evaporation changes our drop from a liquid into a gas or vapor.

On the other hand, our drop might soak into the earth and be taken up by the roots of a plant. Plants return water to the air through their leaves in the form of vapor —an action called **transpiration.** If no plant captures our drop, it might travel downward through rocks and earth. This movement is called **infiltration.** Our drop could work its way down to the groundwater. This process, known as **recharge,** refills the underground water supply.

If our drop does not evaporate quickly or nourish a plant or soak deeply into the

As part of the water cycle, rain collects and runs downhill until it reaches a river or lake. These rapids in the Central African Republic quicken a river's pace as it flows through the western part of the country.

earth, the water may run downhill until it enters a river or a lake. This water is known as **runoff.** From there, the drop may continue all the way to an ocean. The sun's heat turns some ocean water into water vapor, which goes through the air and gathers in storm clouds. Once our drop goes back to the clouds, the water cycle starts all over again.

Not all water beneath the surface of the earth is called groundwater. Water officially becomes groundwater when it reaches the **water table**—the level below which the soil or rock is completely filled, or saturated, with water. The area above the water table is the **unsaturated zone,** and the area beneath it is the **saturated zone.**

CHAPTER TWO

GROUNDWATER'S HIDING PLACES

The individual sections of rock and soil that store groundwater are called **aquifers** (meaning "water-carriers"). They are separated from each other by layers of **impermeable rock**—rock with extremely low permeability that keeps the water in the aquifer.

Aquifers vary a lot. Some are narrow strips. Others cover hundreds of thousands of square miles of underground area. These storehouses of fresh water may be a few feet or hundreds of feet thick. They may lie just beneath the earth's surface, or they may exist in places hundreds of feet deep.

The water in aquifers may be new, that is, recently refilled from rainfall or other moisture that went through the water cycle. Or the water may be very old—a type known as **fossil water** that has been trapped in the earth for a long time.

(Left) **In Saudi Arabia, underground fossil water allows people in this dry Middle Eastern nation to grow food crops in the desert. (Above) Water emerges from hiding at a cave opening in southeastern Minnesota.**

17

Hydrogeologists from the United States and Senegal test the depth of a well being dug in this West African country.

Hydrogeologists—people who study underground water—have found, for example, that groundwater beneath Africa's Sahara Desert is thousands of years old. Some aquifers are so ancient the weather conditions that originally formed the aquifers do not exist anymore. As a result, no more water is entering these underground storehouses. When people use up an aquifer's fossil water, that aquifer will be empty.

Aquifers can be either **confined** (bordered on the top and bottom by impermeable rock) or **unconfined** (covered on the top by the water table). Because confined aquifers are trapped between two layers of impermeable rock, the water within them is under pressure. When a well, or hole, is drilled through the top layer of rock, this trapped water can flow with great force to the surface.

Confined aquifers are also known as **artesian aquifers,** after Artois, the region of France where they were first studied. The largest artesian aquifer in the world is Australia's Great Artesian Basin. It extends beneath an area of about 676,000 square miles (1.75 million square kilometers).

Well diggers also drill down to unconfined aquifers. Unlike artesian aquifers,

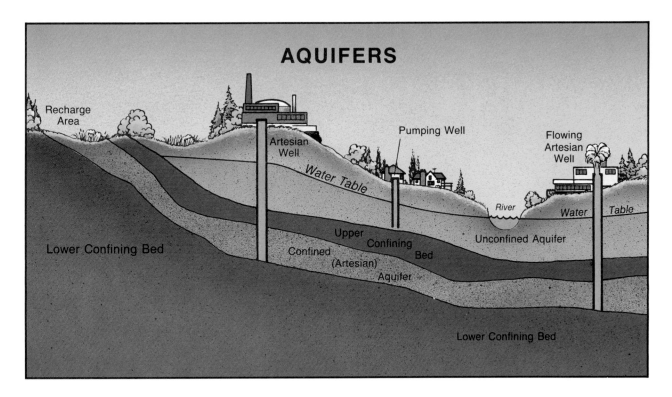

AQUIFERS

Recharge Area

Artesian Well

Pumping Well

Flowing Artesian Well

Water Table

River

Water Table

Upper

Confined

Confining Bed

Unconfined Aquifer

Lower Confining Bed

(Artesian)

Aquifer

Lower Confining Bed

however, the water in an unconfined layer does not come to the surface under its own power. Machines, such as pumps, are needed to bring up this water.

As time passes, groundwater does not stay in one place. It creeps through the aquifer much as a river flows above the ground, only a lot more slowly. "Fast" groundwater may move a few feet in a day. In many cases, groundwater will travel just a fraction of an inch in an entire year. Centuries may pass before the water makes its way from the spot where it seeped into the aquifer to a **discharge area** (a place where groundwater naturally enters a lake or stream).

Discharge areas bring groundwater to the surface, sometimes in spectacular ways. You may have seen springs where clear, cold water comes bubbling up from beneath the earth. **Geysers,** such as Old Faithful at Yellowstone National Park in the United States, are examples of groundwater emerging from hiding.

An oasis (moist, fertile area) in Azraq, Jordan, is surrounded by 12,000 square miles (31,080 square kilometers) of desert.

A geyser erupts with a burst of steam and heated water on Iceland, an island located midway between Greenland and Norway.

So is **quicksand.** It forms when underground water prevents the sand particles from becoming firm enough to support weight. A person who steps into a bed of quicksand can sink very fast into the loose, weak ground.

Oases—isolated fertile areas in an otherwise dry desert—are places where groundwater meets the surface. Water towers and windmills often mark the spot where people have put wells to tap the water supply beneath the ground.

WATER FACTS

- Water is the most common substance found on our planet.

- All living things—plants, animals, and people—need water to live.

- Water is the only natural substance to appear in all three physical forms—solid (ice), liquid (water), and gas (vapor)—at ordinary temperatures.

- Water covers about 70 percent of the earth's surface.

- Although water often changes form and location, the total amount of water on our planet never increases or decreases.

- Humans can live longer without food than without water.

- About 97 percent of the total water on our planet is salty. This salt water lies in oceans and seas. The remaining 3 percent of the earth's total water is fresh water found in ice, snow, rivers, lakes, and the ground.

- Humans and elephants have at least one thing in common—about 70 percent of their bodies are made up of water.

Two mahouts (elephant keepers) wash their elephant in a river in Sri Lanka, an island in the Indian Ocean.

CHAPTER THREE

USING GROUNDWATER

Since ancient times, people all over the world have relied on the water beneath the ground. The Chinese developed the earliest system for drilling wells to depths of several thousand feet. In Cyprus, an island in the eastern Mediterranean Sea, water was once drawn from wells using a chain of buckets pulled by work animals.

Some early wells were so wide that they had special ramps to allow people or work animals to walk down to the water level. About 2,500 years ago, workers in the Middle East dug sloping tunnels that led water from mountain aquifers to villages in the

(Left) **Women in Ethiopia, eastern Africa, help each other draw water at a local well.** *(Right)* **In the 1800s, an explorer from the United States created this picture of a steep, wooden ramp leading down to a groundwater supply in Mexico's Yucatán Peninsula.**

A Saudi Arabian farmer adjusts the water flow from pipes to his crop of corn.

lowlands. These underground channels form one of the oldest public water systems ever recorded.

Before the 1800s, ways of tapping groundwater were limited mostly to springs and shallow wells. Since then, however, people have found better methods for drilling wells and new power sources for drawing water up from the ground. As a result of this technology, the use of groundwater has steadily increased.

In many developing countries, wells provide most of the readily available water. People in parts of the Middle East depend completely on groundwater for their fresh water supply. In Great Britain, about 30 percent of all the fresh water in use comes from under the ground. More than 75 percent of the cities in the United States depend on groundwater for at least some of their water, and 95 percent of rural U.S. homes rely on wells.

In some areas of the world, such as in Japan, groundwater plays a major role in making all sorts of manufactured items.

WATER'S MANY USES

Human beings employ water in many ways. We drink it, bathe in it, and use it to carry away wastes. It is necessary for growing crops and raising animals for food. Without water, we could not make even basic products—like metal, paper, and plastic—let alone cars, houses, and computers.

In the United States, more than 60 percent of the groundwater in use irrigates crops. About 18 percent supplies homes, 13 percent fuels industry, and 1 percent waters livestock.

In India, the irrigation of 40 million acres (16 million hectares) of cropland accounts for 90 percent of the groundwater used there. Most African countries, on the other hand, rely on groundwater mainly for home and industrial purposes rather than for irrigation.

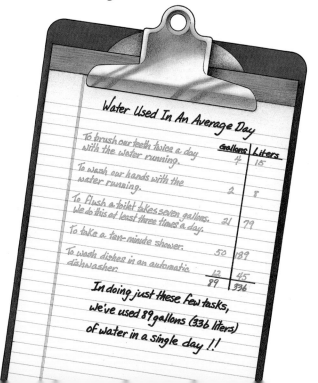

Water Used In An Average Day

	Gallons	Liters
To brush our teeth twice a day with the water running.	4	15
To wash our hands with the water running.	2	8
To flush a toilet takes seven gallons. We do this at least three times a day.	21	79
To take a ten-minute shower.	50	189
To wash dishes in an automatic dishwasher.	12	45
	89	336

In doing just these few tasks, we've used 89 gallons (336 liters) of water in a single day !!

WATER EVERYWHERE?

Before we use groundwater, of course, we have to find it. Some people think that, if you dig deeply, you will eventually hit water. But the precious liquid may not be close enough to the surface or clean enough or plentiful enough to make digging for it worth the trouble.

Since ancient times, people seeking water from the ground have used clues on the surface to guide them. Natural springs—where the water in an unconfined aquifer comes to the surface on its own—are the most obvious signs of groundwater.

If there are no springs, however, the task is to find a promising spot for a well. In general, groundwater is closer to the surface at the bottoms of hills and near lakes and streams. Plants growing in an otherwise dry, barren area provide another hint.

In modern times, advanced technology helps in the search for usable groundwater. Hydrogeologists employ their knowledge of rocks and land, as well as photos taken from the air, to identify likely spots. Scientists hook up special devices that measure

By seeking out groundwater with its deep roots, a tamarisk tree is able to thrive in a Middle Eastern desert.

how electricity travels through rocks. These specialists also record the speed with which shock waves go through the earth. These tests help hydrogeologists to find rocks and other formations that could hold water.

WATER DOWSING
Science or Witchcraft?

Water dowsing is a method of finding underground water. Also called water divining or water witching, dowsing has a long history. Ancient cave paintings in northern Africa depict a **water dowser** at work. Chinese writings suggest that water dowsers were active thousands of years ago. In the 1700s, some people in the United States suspected water dowsers of being witches. Many modern scientists do not accept dowsers as experts at finding groundwater.

Dowsers, who are also known as water diviners or water witches, still practice their craft. They generally use a forked wooden stick that comes from a peach, apple, willow, or witch-hazel tree. The forked stick, or **divining rod,** is usually shaped like the letter Y. Dowsers hold the two prongs in their hands, pointing the shaft of the stick at the sky. When the dowser senses water beneath the ground, the shaft will be drawn downward.

Practitioners of dowsing usually work in rural regions. In urban areas, scientists use computers and other technology to study the ground for possible well sites. Sometimes, however, experts have dug many test wells without finding water. Well-respected dowsers have been called in, and they have surprised the experts by locating water.

Scientists disagree about whether or not water dowsing is a practical way to find water. Many hydrogeologists point out that dowsers are unable to reveal information about water quantity, water quality, and the long-term effects of withdrawal.

A water dowser holds a divining rod.

UP FROM THE GROUND

After the scientists have found a usable aquifer, the next task is to get the water out of the ground to the place where people need it. A natural spring solves the problem easily. Otherwise, workers must drill down to the water. Hand-dug wells also can be used to reach water in shallow aquifers.

Water towers or windmills are above-ground symbols of deep groundwater tapping. Workers drill wells through soil and rock until they reach an aquifer. Modern wells are commonly 1 to 36 inches (2.5 to 91 centimeters) in diameter and are usually lined with a metal or plastic **casing.** This

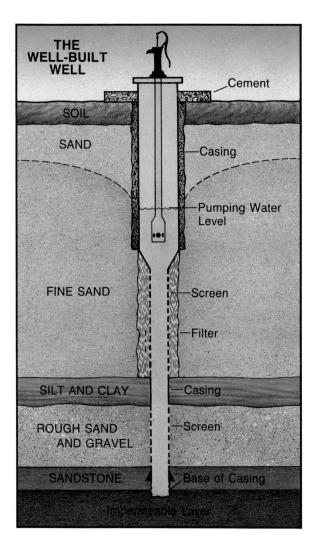

New drilling machinery (left) and modern wells (above) **allow more groundwater to be tapped.**

protection helps to keep anything but water from entering or leaving the well.

If a well taps into an artesian aquifer, water may gush to the surface. Otherwise, energy is needed to draw the water upward. This energy may come from humans who turn a crank or pump the water using a long handle. The work may involve animals harnessed to a mechanical pump. Windmills use the power of the wind to draw up water. In many instances, electric motors pull the water from a well to the surface.

Once it reaches the surface, groundwater may be put into action immediately, or it

By walking in a circle while harnessed to a water-wheel, a camel in Egypt provides the energy, or power, for drawing water from the ground.

may be stored. Water towers are one example of a **reservoir**—a place of storage—for groundwater used by communities. Water is pumped into the reservoir and then flows through a network of pipes to individual homes and businesses.

Water towers dot the landscape of the central United States, where groundwater is the main source of water.

When groundwater is removed from a well, the water table near the well drops—an action called **drawdown.** Well drillers need to keep drawdown in mind when they decide how deep to make a well. If the well is too shallow—if its bottom is too close to the water table—drawdown could make the well go dry even though the aquifer still has plenty of water.

CHAPTER FOUR

TO TAP OR NOT TO TAP

In many ways, groundwater is the ideal source of water. For one thing, there is a lot of it. In fact, aquifers near the earth's surface hold much more water than all of the rivers in the world combined. Also, groundwater is found around the globe, even in places where surface water is scarce. And groundwater does not have to be caught like rain or dammed like a river. Instead, groundwater can be withdrawn as it is needed.

Because groundwater lies below the earth's surface, it comes with a built-in storage room. This resource does not take up large areas of land like above-ground reservoirs do. In California's Central Valley, for example, the top 100 feet (30 meters) of earth can hold more water than all of the state's surface-water reservoirs put together.

(Left) **Farms in the central United States are big users of groundwater.** *(Above)* **In some parts of Africa, people climb down into the ground to pass up small supplies of water.**

One of the biggest advantages of groundwater is that it is usually fairly clean. Groundwater travels through soil, sand, and rock. These materials filter out some of the harmful organisms and chemicals—called **pollutants**—that make the water unusable. This self-cleaning feature is important in the world's poorest countries, where systems to purify drinking water are not always available.

NOBODY'S PERFECT!

But groundwater has its disadvantages, too. In some areas, tapping groundwater may be too hard or too expensive. For ex-

A farmer in Burma carefully waters her crop of jasmine, a flower that is popular in her Asian country.

Workers drill a family's well by driving a pointed metal pipe into the ground until it reaches the water table.

ample, there is a lot of groundwater from melting snow in Alaska, but the ground there is permanently frozen below a certain depth. This locks in the water, making very little of it available.

In countries where groundwater supplies lie far beneath the earth, experts need expensive tools and machinery to tap the aquifers. Many poor nations do not have the technical knowledge and equipment to reach their groundwater.

Sometimes groundwater contains small amounts of minerals—such as calcium, magnesium, and iron—that affect the water's taste and usefulness. This water is commonly called **hard water.**

The surface water that trickles down through the soil contains carbon dioxide, a gas found in rain. The mixture of carbon dioxide and water forms a very weak acid. Over time, the weak acid in groundwater dissolves the minerals in the rocks. The minerals stay in the water and give some groundwater a distinct taste and smell. Some people use water-softening chemicals to remove the dissolved minerals.

CHAPTER FIVE

GROUNDWATER AROUND THE WORLD

From watering camels in the Sahara Desert to cooling off Little-League ballplayers in South Dakota, groundwater has a lot of jobs to do. It played a part in producing the paper for this book. You may have used groundwater when you brushed your teeth this morning. Right this minute, groundwater is at work in thousands of places all over the globe.

NEBRASKA'S HERO

After school, Kate Anthony watches part of her family's flock of sheep graze and drink water. As she walks to her house in Nebraska, she looks across the distant wheat fields at the center pivot irrigator—a long, rotating sprinkler that draws water from a well. Although Kate probably does not know it, her family's healthy livestock

and rich cropland depend on the huge Ogallala Aquifer far beneath her feet.

This aquifer is the unseen hero of the central United States. It provides water for about 170,000 wells in Texas, Oklahoma, New Mexico, Kansas, Colorado, Nebraska,

(Left) **Boys in the Asian kingdom of Bhutan bathe under a flowing water pipe. (Above) A center pivot irrigator sprays water in a wide curve over a Nebraska field. The water comes from the Ogallala Aquifer, which also supplies several other states in the central United States.**

Wyoming, and South Dakota. The Ogallala covers an area of about 174,000 square miles (450,660 square kilometers). It is about 450 feet (137 meters) thick, making it one of the largest aquifers in the United States.

The aquifer is very important to the people in the area. The Ogallala's water irrigates 143 million acres (58 million hectares) of land. Without this aquifer, life would be much different in a region that is known for its huge herds of livestock and waving expanses of grain. With the Ogallala's water, people like Kate and her family can grow crops and raise animals to feed thousands of people.

HOT TIMES IN ICELAND

Jon Jonsson does not have a furnace in his house in Reykjavík, the capital of Iceland. He does not need one, either—even though temperatures in his country average 30° F (–1° C) in the middle of winter.

The earth's crust is thin in Iceland, an island in the North Atlantic Ocean, where underground volcanic activity is common. As a result, the groundwater there is naturally hot and close to the surface.

The hot water is pumped through pipelines that lead from the city's wells directly into Jon's home. The water circulates through radiators, giving off heat that keeps Jon and his family warm in the coldest weather. In fact, hot groundwater fuels the heating systems in just about every house in Reykjavík.

Most of the homes in Reykjavík, Iceland, are warmed with naturally heated groundwater.

THE STEAMING EARTH
or How Geothermal Energy Works

A hot spring in Wyoming's Yellowstone National Park is a sign of underground geothermal energy at work.

Geothermal energy comes from the natural heat that lies at the core, or center, of our earth. Temperatures at the core are extremely high, sometimes reaching thousands of degrees Fahrenheit. Far away from the core—at the earth's crust, or surface—temperatures are much lower.

In volcanic areas, however, the crust stays very hot and contains melted rock, called **magma.** The rocks that surround magma are also hot, and their pores hold naturally heated groundwater. In parts of Iceland, New Zealand, Indonesia, and the United States, heated groundwater produces some unusual formations above ground. These include hot springs, mud pots, and geysers.

Iceland captures much of its hot groundwater in underground pipes. These redirect the water to heat homes, public buildings, and even greenhouses.

Other countries use their naturally heated groundwater to produce electricity in geothermal plants. At these plants, engineers have drilled wells to bring up hot water and steam—energy forces that turn huge wheels, called turbines. The turbines drive generators that produce electricity, which is sent to homes and businesses near the plant.

TREKS IN WEST AFRICA

Five gallons of water weigh half as much as she does, but Fanta Ndaye is used to carrying the load. Part of her daily responsibility is providing water to her family, even if the water source is miles away.

Twenty years ago, the people in Fanta's village in Senegal, West Africa, drew water from hand-dug wells not far from their homes. Fresh water would seep back into the shallow holes overnight, ready for the next day's use. In recent years, however, the nearby Sahara Desert has been spreading, and many of Senegal's hand-dug wells have gone dry.

The people in Fanta's community now depend on the deeper Maestrichtian Aquifer for their water. Lying several hundred feet below the surface, this aquifer can only be reached by machine-driven wells. Because well-drilling equipment is scarce in Senegal, so are wells. Fanta's village no longer has its own well. As a result, Fanta and the other girls and women frequently make their trek for water so that their families can live.

A young Senegalese girl totes water back to her family's village from a community well.

JAPAN'S SINKING FEELING

Yoshi Yamamoto is thinking about groundwater. He has just read in the newspaper that parts of eastern Tokyo, the capital of Japan, are sinking. The groundwater beneath the city is being withdrawn for use in the area's industries. This water was supporting the land above it. Where, Yoshi wonders, is all the water going?

To answer that question, Yoshi need not look much farther than his mirror. Along with the other 8.3 million people who live in Tokyo, he is doing his share to consume the goods that require water to be produced. The radio on his dresser was made using groundwater. The soda he stopped to drink on the way home from school contained it. Groundwater helped to print the

Parts of Tokyo are sinking because of groundwater withdrawal.

newspaper with the story that caught his eye. Simply by living in Tokyo and by using things manufactured there, Yoshi is contributing to that "sinking feeling."

WELLS IN LLAMA LAND

To ward off the chilly air, Elena Garcia pulls her sweater more tightly around her. She lives in the northwestern part of Peru, a South American nation that is famous for its camel-like llamas.

During the summer months, the Chicama River rushes past the small plot of land that Elena and her family farm near the Pacific coast. The river is fed by melting snows that flow westward from Peru's Andes Mountains. In winter, very little rain falls in Elena's area, and only a trickle of water marks the Chicama's riverbed.

When the surface water disappears, Elena and her family must rely on one of the 15,000 wells that dot the Peruvian countryside. These wells tap shallow groundwater, supplying water to people and animals that live along the coast.

CHAPTER SIX

OUR ENDANGERED GROUNDWATER

Groundwater is important in the everyday lives of people all over the world. But this resource is also in danger. In some places, people are tapping groundwater faster than nature can replace it. In other areas, it is being carelessly dirtied with pollutants. Some countries have covered the land with so much pavement that water cannot enter the soil and recharge underground supplies.

USING IT UP

In the late 1800s, families in North and South Dakota searched for a reliable water source. As they dug into the Dakota sandstone, they hit a large aquifer. Wells in this aquifer sent water spraying as high as 100 feet (30 meters) into the air. The local people now had fresh, clean water as well as

(Left) Pavement covers parking lots and highways and prevents water from entering underground aquifers. (Above) An old photograph shows one of the free-flowing wells that settlers dug into the Dakota Sandstone Aquifer in the late 1800s.

a strong energy source. The pressurized water powered machinery, watered live-stock, and irrigated farmland.

As workers drilled more and more wells, however, the pressure in the first wells weakened. By the 1920s, engineers had driven more than 16,000 wells into the Dakota Sandstone Aquifer. Rather than using groundwater as a renewable resource, the people were exhausting it by taking water out faster than the water cycle could replace it.

Such draining, called **overdraft** or **mining,** is one of the most serious threats to groundwater. Overdraft is occurring around the world. Communities in West Africa often face drought and must rely more and more on underground water. As a result, their aquifers are running dry.

Israel and a neighboring Middle Eastern area called the West Bank both use groundwater. In the 1970s and 1980s, the water level of an aquifer shared by Israel and the West Bank fell when Israel increased its use of groundwater. In North America, farmers are drawing water from

Sinkholes sometimes form where too much groundwater has been withdrawn, causing the surrounding rock to collapse inward.

the Ogallala Aquifer much faster than it is being recharged. Water levels in this aquifer are dropping one to five feet every year.

Overdraft also causes **subsidence** (sinking) of the ground. In some places, such as Tokyo, the water in the aquifer actually supports the earth above it. As the water level drops, so does the land surface. In Bangkok, the capital of Thailand, the land is sinking roughly five inches (13 centimeters) each year because the groundwater there is being removed too quickly.

Another problem related to overdraft is **salt-water intrusion**—the entry of sea water into an aquifer. When fresh water is pumped from an aquifer that lies near the ocean, salt water is often pulled in to replace the tapped groundwater. The Netherlands, Israel, and Florida are three places where salt-water intrusion is contaminating aquifers.

WASTE NOT, WANT NOT

Some people believe that overdraft is just the price we pay for tapping groundwater. But this valuable resource is not being employed wisely. In the United States, the amount of groundwater used for irrigation tripled between 1950 and 1980 to 60 billion gallons (227 billion liters) per day.

Of course, this resource helped to feed a growing population. But we might have used less groundwater if we had realized the supply was limited.

In many places, irrigators flood fields. This method consumes more than twice as much water as is needed to nourish plants.

These sprinklers in a Turkish field are a less wasteful way to irrigate farmland than flooding.

Before it can be cleaned up, polluted surface water can seep down into aquifers and endanger groundwater.

Other ways of distributing the water—such as sprinklers or **drip irrigation**—let the plants directly absorb up to 90 percent of the water used to irrigate them.

The French government found another successful way to reduce water waste. Since 1969, the government agencies that control the French water supply have charged a fee for water. Because the fee is based on the amount of water used, people have cut back on their water consumption. The government uses the money to pay for conservation and water-treatment projects.

POLLUTED GROUNDWATER

Lying far beneath layers of soil and rock, groundwater seems safe from the pollution that threatens air, surface water, and land. Although these layers do protect groundwater in some cases, many pollutants enter aquifers.

Contaminants reach groundwater through old or poorly lined wells or through cracks in rock layers. Polluted lakes and streams that connect with groundwater also endanger it. Because groundwater is a hidden resource, often nobody notices that it has been damaged until the pollution has spread, making cleanup difficult or impossible.

Aquifers can be polluted in many ways. Accidents such as chemical spills are very

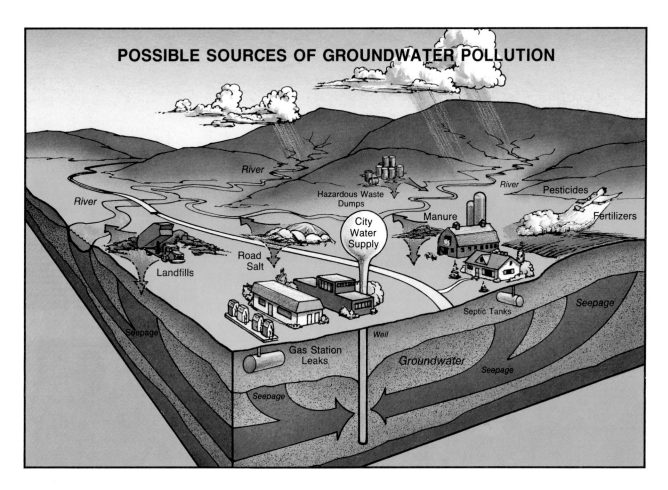

POSSIBLE SOURCES OF GROUNDWATER POLLUTION

River

River

River

Hazardous Waste Dumps

Pesticides

Manure

Fertilizers

City Water Supply

Road Salt

Landfills

Septic Tanks

Seepage

Seepage

Well

Gas Station Leaks

Groundwater

Seepage

Seepage

obvious sources of pollution. For example, in 1974 fire fighters in a town in Virginia battled a blaze behind a service station. The water they sprayed on the fire washed used car oil—which had collected for decades behind the station—into the local aquifer. As a result, 18 wells had to be shut down.

Most of the time, however, groundwater pollution results from our day-to-day behavior. Suppose you are a clean, fresh aquifer, untouched by human action. You might shudder at the list of your potential enemies.

SEPTIC TANKS—underground systems that remove pollutants from waste water—can leak chemicals and harmful

An investigator takes soil samples to determine if the groundwater near a river has been polluted.

bacteria (small organisms) into groundwater. In the United States alone, 20 million such systems treat more than 2 *trillion* gallons (7.5 trillion liters) of water each year. So even a small number of leaky septic tanks can spell big problems for groundwater.

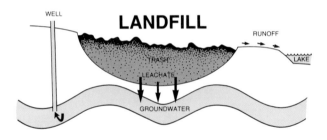

LEACHATE, a sort of "trash soup" made of liquids and garbage, can travel from **landfills** (garbage dumps covered with soil) into groundwater. The leachate can carry germs and chemicals from garbage into groundwater. In West Germany, about 35,000 landfills threaten groundwater. Newer waste-disposal sites often are built with special liners to protect the water below the land. Even these liners occasionally spring leaks, however.

Meet TRACY ADAMS...

Tracy lives in Hopkinton, Massachusetts, not far from the city of Boston. In 1989 she won one of the U.S. president's Environmental Youth Awards. Only 10 of these awards are given each year.

What did Tracy do to earn this honor? During her junior year at Hopkinton High School, she started a program to make people aware of hazardous household chemicals, which are found in bleach, oven cleaner, and paint. The cleaners and other strong substances we use in our homes can pollute groundwater if we do not get rid of the products and their containers in the right way.

Tracy designed a display and wrote a report to educate students and their families about how to dispose of dangerous household substances without hurting the environment. She also drew up a list of alternative, non-hazardous products.

Through Tracy's work, Hopkinton and near by communities held a hazardous-waste collection day. People brought their old paint, used car oil, and other chemicals to the collection area. Special disposal methods and recycling efforts prevented these substances and their containers from getting into landfills or groundwater.

On large farms, planes often spread pesticides and fertilizers over a wide area. These substances can end up in groundwater.

PESTICIDES and FERTILIZERS that farmers and homeowners spread on cropland and lawns can move down into aquifers. These chemicals can make water unsafe for drinking. In many countries, aquifers that recharge from farmlands contain large amounts of **nitrate,** a form of nitrogen that comes from fertilizers. Nitrate can cause health problems in humans and livestock. In the United States, 50 different kinds of pesticides have been found in aquifers in 30 states.

THE TRAGEDY OF LOVE CANAL

In the late 1800s, a land developer named William T. Love got money to dig a canal to connect the upper and lower parts of the Niagara River. By 1910 Love had lost his funding, and the partially dug canal was abandoned.

For a while, the trench-like hole—called Love Canal—became a community swimming pool for the city of Niagara Falls, New York. In 1947 the Hooker Electrochemical Corporation bought Love Canal to use as a dumping site for dangerous toxic wastes. Between 1947 and 1953, the Hooker company put about 20,000 tons [18,144 metric tons] of harmful chemicals into the canal.

In 1953 the corporation filled the rest of the canal with dirt, capped it with clay, and sold it to the Niagara Falls Board of Education. The sales agreement stated that the company was not responsible if anyone got hurt or died from the buried wastes. Homes and a school were built on and around the canal. For decades, families lived in the neighborhood unaware of the potential dangers beneath the ground.

In the mid-1970s, after heavy rains, the buried chemicals began to seep up through the soil and clay, polluting the groundwater and soil of Love Canal. A thick, black sludge appeared in the basements and on the lawns of homes. Eventually, homeowners realized that children in Love Canal frequently developed diseases and that pregnant women in the neighborhood sometimes lost their babies. People wondered if these events had a connection to the sludge.

Worried citizens asked the state of New York for help. Scientists tested the water and soil and found them polluted with the buried chemicals. In 1978 New York authorities evacuated all pregnant women and all children under the age of two. The state also bought the houses bordering the canal and closed the school. But the sludge continued to move beneath the ground. In 1980 President Jimmy Carter issued an order to move all families in Love Canal away from the abandoned dump site. Ten years later, officials stated that some parts of Love Canal were safe, and some abandoned houses are up for sale. The U.S. government says the nation has about 50,000 toxic-waste dumps just like Love Canal.

At hazardous waste dumps, trained experts dispose of the dangerous chemicals we use in our homes and industries.

INDUSTRIAL WASTE may damage groundwater if the waste is not handled carefully. Years ago, people dumped many industrial chemicals into groundwater. In the early 1900s, for example, a manufacturer in Great Britain disposed of a dangerous chemical—picric acid—in abandoned chalk pits. Within 10 years, the acid showed up in nearby wells. By 1955, it had traveled a mile through the aquifer. Another 60 years may pass before the aquifer is completely free of the contaminant.

SALTS, which are spread on highways to melt snow during the winter, enter and pollute groundwater. Almost 10 million tons of salt are used in this way each year in the United States. In the spring, the salt and melting ice mix together and seep

GROUNDWATER GUZZLERS

Although people and their industries use lots of groundwater, some plant species—called **phreatophytes**—are also groundwater guzzlers. The roots of these plants reach down deeply to find water, going below the water table into groundwater. Among the most common phreatophytes are tamarisks, mesquites, and cottonwood trees.

Between 1899 and 1915, the U.S. Department of Agriculture imported eight species of tamarisk, also known as salt cedars, to the southwestern United States. These plants require immense quantities of groundwater to survive. In fact, tamarisks in the western United States use about 20 trillion gallons (75.7 trillion liters) of groundwater annually.

People have learned to let phreatophytes show them where to dig wells. For example, mesquites survive only where a permanent water supply exists within 30 feet (9 meters) of the surface. Many wells in California's desert have been dug between mesquite shrubs.

Cottonwood trees thrive on groundwater in Utah's desert.

In winter, trucks spread salt to de-ice the highways. The salt helps to melt the ice and improve driving conditions.

into groundwater. Communities in Massachusetts had to stop using several wells that had become contaminated by salt.

INJECTION WELLS, extremely deep wells, are used to dispose of wastes. In 1962, for example, workers drilled a 12,000-foot (3,657-meter) well in Colorado to bury liquids used in making chemical-warfare weapons. Such wells do not often contaminate groundwater because they are

so deep. But at this Rocky Mountain Arsenal, the chemical wastes escaped through cracks in the well casing.

CREEPING CRUD

Polluted groundwater would be relatively simple to deal with if it stayed in one place. Unfortunately, it spreads through the aquifer in a process called **dispersion.**

How far and how fast the pollution extends depends on the pollutant and the aquifer. Materials, such as oil, that do not dissolve in water will spread in a thin layer. Harmful substances that are heavier than water will drop to the bottom of the aquifer.

The many different ways in which pollutants move through groundwater can make their sources very hard to identify. In addition, specialists cannot always tell how far a pollutant has spread. Pollution-control experts often have to install many special wells, called **monitoring wells,** to track the damage to groundwater supplies.

NO QUICK FIXES

The cleanup of polluted groundwater is almost always difficult and expensive. Barriers are sometimes built to keep the contaminants from spreading farther. Some pollutants can be destroyed right in the ground. Water also may be pumped from the ground, cleaned up, and re-injected into the aquifer. If pollutants have had a lot of time to move around, they often are extremely hard to surround and clean up.

Whatever technique is used, removing pollutants from groundwater is likely to be more difficult, more time-consuming, and more expensive than keeping out harmful substances in the first place.

Workers in Minnesota use dredges, pipes, and hoses to try to clean up an oil spill.

CHAPTER SEVEN

PROTECTING GROUNDWATER IS EVERYBODY'S JOB

Groundwater is a plentiful and important resource on our planet. But this valuable water supply is in danger of being made unusable through human carelessness and wastefulness. If we want groundwater to continue to work for us, we must do our part to keep it clean and available.

It might be hard for you to imagine being able to make a difference in the fate of such a huge storehouse of water. Taken together, however, the actions of individuals like you have a tremendous impact. Our day-to-day activities *can* contribute to the preservation of this valuable resource.

Here are some things we all can do to help to protect groundwater.

USE WATER CAREFULLY. Each person in the United States consumes up

(Left) A zeppelin brings an air-borne message about water conservation to a city in the United States. (Above) Using an earthenware container, an Ethiopian woman carefully waters a small patch of young trees.

to 80 gallons of water a day. Daily U.S. water consumption is 3 times that of Japan and is more than 70 times what people in Ghana, West Africa, use. All of this means that there is plenty of room for cutting back. If you believe that groundwater is a valuable liquid and not "just water," you may think twice before wasting it.

DON'T LITTER. Never put trash or other sources of pollution in lakes and streams and never bury chemicals in the bare ground.

REDUCE, REUSE, RECYCLE, DO WITHOUT. Think about the trash you create and help others to think about it, too. For example, when you purchase something small, avoid taking it from the store in a bag, especially a plastic one. The less garbage we produce, the fewer land-fills we will need. With fewer landfills, we make it less likely that groundwater will become polluted from our garbage.

HELP YOUR PARENTS TO DIS-POSE OF HOUSEHOLD CHEMICALS PROPERLY. Old paint, used car oil, and leftover cleaning fluids can hurt the ground-

water if they end up in landfills or are dumped on the ground.

BE ON THE LOOKOUT FOR LEAKY FAUCETS. Turn them off or encourage their owners to fix them. That little drip . . . drip . . . drip uses as much as 20 percent of our fresh water supply.

Recycling all possible waste products—aluminum, paper, glass, cardboard, and lawn clippings —saves space in landfills.

HELP ADULTS TO THINK ABOUT CONSERVING WATER. Remind them about the value of water if they run partial loads of laundry or leave the lawn sprinkler on all night. Urge them to consider putting in water-saving toilets and showers.

APPRECIATE NATURE. Point out its beauty to other people. We can save a lot of water if we use native plants, such as cactus or prairie grasses, to brighten our surroundings rather than planting lawns and shrubs that need extra water to stay alive.

REPLACE CHEMICALS WITH MUSCLE POWER. Instead of using fancy foam, clean your sink with some good hard rubbing. Instead of relying on gas-fueled vehicles, ride your bike to the store.

OFFER TO PULL WEEDS INSTEAD OF USING CHEMICALS ON LAWNS AND GARDENS. If you know people who use lawn fertilizer, ask them if they are sure that they are using only as much as they need. Help them to see what happens to the extra fertilizer that stays in the ground.

REPORT WATER DANGERS. If you notice or hear about an abandoned well, report it to your local health department or environmental organization. These holes can bring pollutants quickly to groundwater.

AMERICA'S CLEAN WATER FOUNDATION
Hall of the States
444 North Capitol Street NW, Suite 330
Washington, D.C. 20001

AMERICAN WATER WORKS ASSOCIATION
6666 West Quincy Avenue
Denver, Colorado 80235

CLEAN WATER ACTION PROJECT
317 Pennsylvania Avenue
Washington, D.C. 20003

CONCERN, INC.
1794 Columbia Road NW
Washington, D.C. 20009

LEAGUE OF WOMEN VOTERS
1730 M Street NW
Washington, D.C. 20036

SOIL AND WATER CONSERVATION SOCIETY
7515 Northeast Ankeny Road
Ankeny, Iowa 50021

WORLDWATCH INSTITUTE
1776 Massachusetts Avenue NW
Washington, D.C. 20036

U.S. GEOLOGICAL SURVEY
Hydrologic Information Unit
MS 419 National Center
Reston, Virginia 22092

Photo Acknowledgments

Photographs are used courtesy of: p. 4, NASA; p. 6, USDA Forest Service; p. 7, Jay A. Stravers; p. 8 (left), Leslie Fagre; p. 8 (right), Eleanore Woollard; p. 9 (left), Amoco; p. 9 (right), 10, 27, 29 (right), 30, 35, 42, 48, 49, 59, 60, National Association of Conservation Districts; p. 11, American Association of Petroleum Geologists; p. 13 (all), 26, Independent Picture Service; p. 14, 39, 54, Minneapolis Public Library and Information Center; p. 15, Baptist Mid-Missions; p. 16, Isabel Cutler; p. 17, 18, 33, 38, 41, 50, 53, 54, 62 (right), Hans-Olaf Pfannkuch; p. 20 (left) B. F. Molnia; p. 20 (right), Jordan Information Bureau, Washington, D.C.; p. 21, (right) Nathan Rabe; p. 22, Ray Witlin/World Bank; p. 23, Boston Athenaeum; p. 24, 34, F. Mattioli/FAO; p. 25 (left), Mitsubishi Motors Corporation; p. 28 (left), 62 (left), George E. Failing Company; p. 29 (left), Drs. A. A. M. van der Heyden; p. 31, Phil Porter; p. 32, 43, F. Botts/FAO; p. 36–37, Leonard Soroka; p. 37, David Falconer; p. 40, 52, Minnesota Department of Transportation; p. 44, Cindy Turkle/Iowa Geological Survey Bureau; p. 46 (left), U.S. Environmental Protection Agency, Region 5/Lorna Jereza; p. 47 (left), Minnesota Pollution Control Agency; p. 47 (right), Tracy Adams; p. 51, Elizabeth Pilon; p. 55, World Bank; p. 56–57, Jerry Boucher. Charts and illustrations by Bryan Liedahl.

Front Cover: American Cave Conservation Association
Back Cover: (left) Yosef Hadar/World Bank; (right) Hans-Olaf Pfannkuch

aquifer (AK-wi-fer): a layer of rock or sediment containing groundwater that can be drawn out for use above ground.

artesian (ar-TEE-zhun) well: a well that is drilled into a confined aquifer. The water in this type of aquifer may gush to the surface without the need for a pump.

bacteria (bak-TEER-i-yah): groups of very small organisms (micro-organisms) that eat living or dead materials.

carbon dioxide: a gas that is naturally found in the air and combines with water to form carbonic acid.

casing: the protective lining of a well.

confined aquifer: an aquifer that is trapped between two impermeable layers of material.

discharge area: a place where groundwater leaves an aquifer.

drawdown: the lowering of the water table that occurs near an active well.

dispersion: the movement of pollutants through an aquifer.

drip irrigation: a direct, water-saving method of nourishing plants. Drip systems trickle water to plant roots from small holes in pipes.

evaporate: to change water from a liquid to a gas (vapor).

fertilizer: a natural or chemical substance added to the soil to help plants grow.

fossil water: water in an aquifer that is no longer being refilled.

geothermal energy: energy produced by the heat of the earth's interior.

Using drip irrigation on farms is one way of conserving precious groundwater.

geyser (GUY-zer): a hot spring through which a jet of groundwater and steam sometimes erupts. Geysers form in areas where there is a lot of volcanic activity.

glacier (GLAY-sher): a huge mass of ice that moves slowly down mountain valleys or over land.

groundwater: water that lies below the water table.

hard water: water containing large amounts of dissolved minerals.

hydrogeologist: a person who studies groundwater. The study of groundwater is called hydrogeology.

impermeable (im-PER-me-abl) rock: rock that almost completely blocks groundwater from traveling through it.

industrial waste: pollutants created by factories that manufacture goods.

infiltration: the process by which water moves into and through soil and rock to an aquifer.

injection well: a very deep well into which liquid wastes are dumped for disposal.

landfill: a garbage dump covered with a thin layer of soil.

Groundwater usage has increased in the Sunbelt, as more people move to this warm, sunny part of the southern and southwestern United States.

leachate: a liquid that forms at the bottom of landfills as water filters through trash.

magma: hot, melted rock.

monitoring well: a well drilled to check the purity of groundwater or to observe the movement of pollution.

nitrate: a chemical compound used in fertilizers that helps plants to grow. When washed into a water supply, nitrates can cause water pollution.

oasis (oh-AY-sis): a place where groundwater comes to the surface in an otherwise dry area.

overdraft: the situation that occurs when more water is drawn from an aquifer than enters it.

permeability: a measure of how easy it is for water to travel through rocks or sediment.

pesticide (PES-ti-side): a chemical used to destroy insects or other pests.

phreatophyte (free-AT-o-fite): a plant that sends roots into groundwater.

pollutant (po-LOOT-int): a chemical or organism that contaminates groundwater.

pores: the small holes within rocks that can hold groundwater.

porosity (po-ROS-i-tee): a measure of the amount of "empty" space in the pores of rocks or sediments.

quicksand: a mass of loose, water-filled material that behaves like a liquid. A heavy object placed on the surface of quicksand sinks quickly.

recharge: the process by which water enters and refills an aquifer.

reservoir (REZ-er-vwar): a place where people store water for later use.

resource: a supply of something we need.

runoff: water that does not soak into the ground but runs into a river, lake, or ocean.

salt: a chemical that is spread on highways and sidewalks to melt snow in winter.

salt-water intrusion: the movement of salt water into an aquifer as the groundwater is removed.

saturated (SATCH-uh-rated) zone: the area below the water table that is completely filled with water.

Drills are often needed to reach groundwater in an unconfined aquifer.

septic tank: an underground sewage container that sifts human waste. The waste stays in the tank until bacteria break down the sewage into products that are flushed into the ground.

subsidence (sub-SIDE-ints): a sinking of the earth's surface that sometimes occurs when groundwater is removed from an aquifer.

transpiration: the movement of water through a plant's roots and leaves and out again into the air.

unconfined aquifer: an aquifer that has the water table as its upper boundary.

unconsolidated (un-cun-SOL-i-dated) sediments: loose materials, such as sand and gravel, that have not been cemented together into rock.

unsaturated zone: the area above the water table that is not entirely filled with water.

water cycle: the various paths and forms that water takes as it circulates through the air to the earth and back again.

water dowsing: a method of finding water underground. Water dowsers use a Y-shaped stick known as a divining rod.

water table: the boundary between the saturated and unsaturated zones.

Sometimes groundwater comes to the surface on its own in the form of boiling springs.